Learning with Findus and Pettson

Telling the Time

Pettson's schedule

Pettson likes to plan his day. Can you match the times on the clocks to the correct activities?

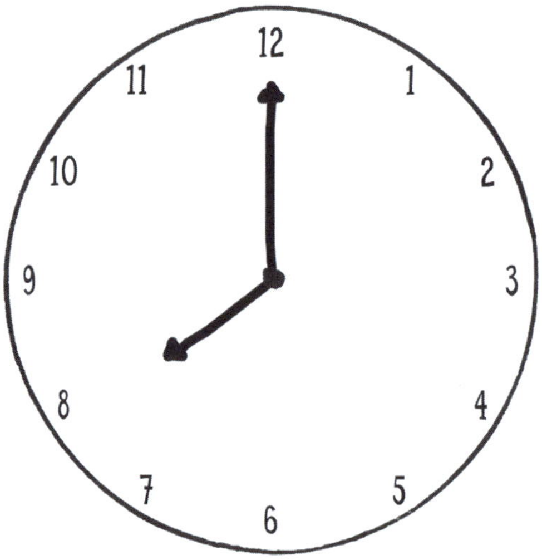

Have breakfast at 8 o'clock

Do the crossword at 11 o'clock

Tinker in the shed at 3 o'clock

Cock-a-doodle-doo!

Pavarotti the rooster's most important job is to wake everyone up at 7 o'clock. But he's not an early bird. What time does he start crowing on the following days?

Monday

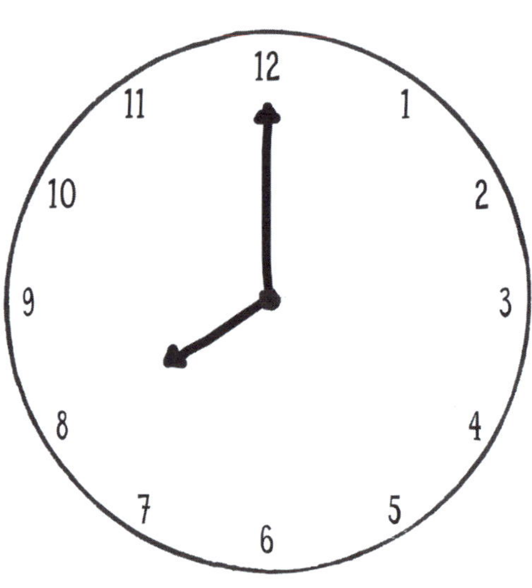

____

Tuesday

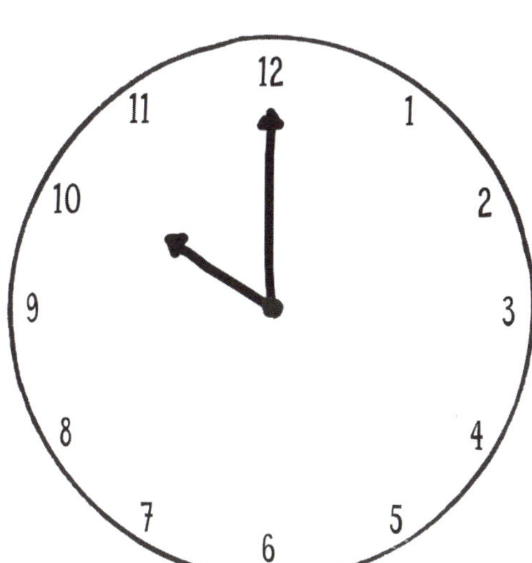

Wednesday

What time is it?

The muckles have been playing in the old clock and now it's stopped working. Can you help them by drawing on the missing hands to show the correct time?

9 o'clock

4 o'clock

2 o'clock

Half past ten

What is Findus doing?

Findus wakes at 5 o’clock every morning, ready for breakfast and play. Would you like to draw what he does at the different times?

Breakfast at 6 o’clock

Jumping on the bed
at 9 o'clock

Fishing at 11 o'clock

Will Findus be late?

It is a quarter to two and Pettson will serve pancakes at 3 o’clock. Findus can choose three different ways to get home. Which path should he take to get back in time?

One and a
quarter hours

The fox is coming!

The fox is on the prowl, so Findus and Pettson must make sure to lock up the hens at night. To do things properly they should take the following steps – but in what order?

Shut and bolt the door

☐

☐

Keep a lookout for the fox

Herd the hens into their coop

Put the hens to bed

Moving house

The muckles are sick of the rooster's crowing, so they decide to move to the other end of the garden. But their cart is too full and things keep falling off. Every hour two things fall off. How long have the muckles been travelling?

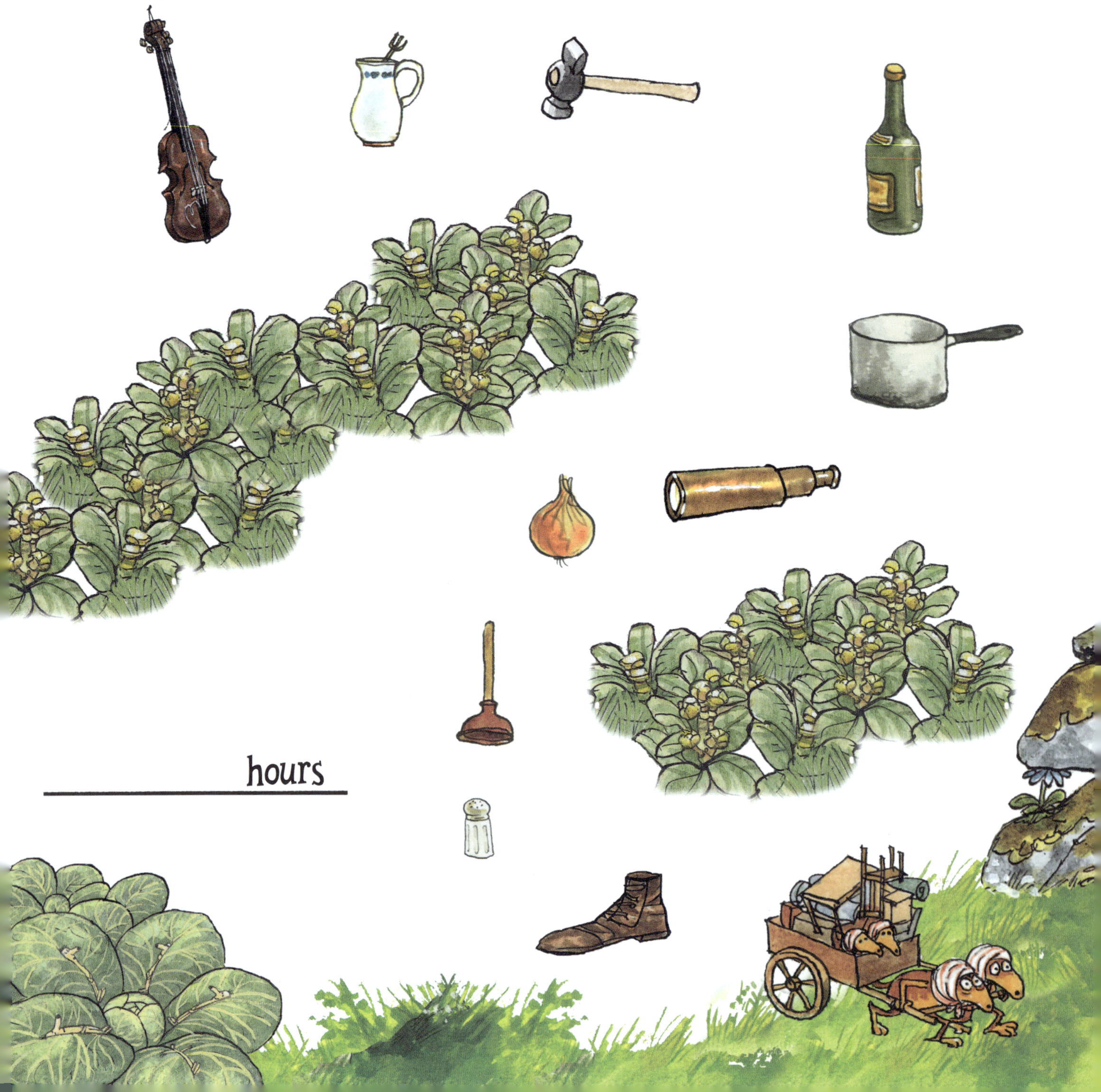

______________ hours

What time is it?

The muckles have been at it again.
Draw the hands on the clocks to show the right times.

7 o'clock

Half past six

How long to go?

It is 5 o'clock in the morning. Findus wonders how many hours are left until the things in the pictures happen. What do you think?

____________________ hours

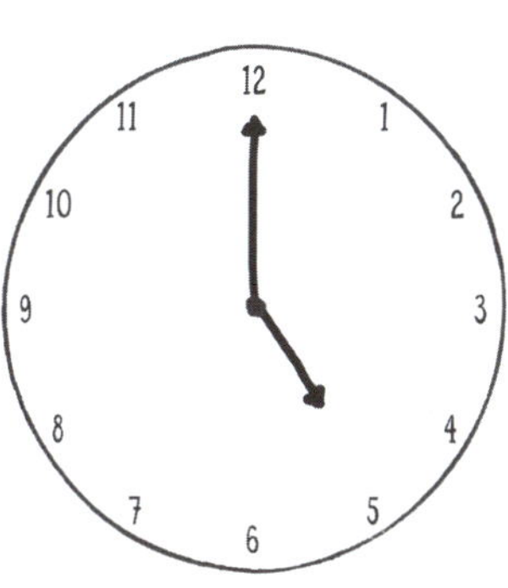

____________________ hours

How long to go?

It is now 2 o'clock in the afternoon. Findus wants to know how long he must wait until the fun starts. Can you help him work it out?

______________________ hours

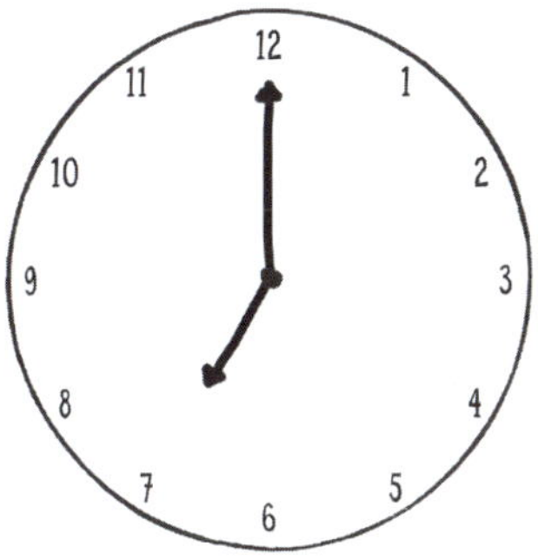

11 12 1 10 2 9 3 8 4 7 6 5

_______________ hours

11 12 1 10 2 9 3 8 4 7 6 5

_______________ hours

Muckle get-together

Two muckles plan to have a picnic under the big tree at 3 o'clock. The muckle in the tall house lives further away and it will take her one hour to reach the tree. It will only take her friend half an hour. What time should each of them leave home to arrive on time?

12
1
2
3
4
5
6
7
8
9
10
11
12
1
2
3
4
5
6
7
8
9
10
11

A walk in the garden

Findus was lost in the garden and it took him an hour and a half to find his way out. Can you draw the path he took to get out and write what time he came out?

He entered here at two o'clock

He came out here. The time was ____________

Party time!

It's Findus's birthday and he's having a party. The artist muckle has helped to make the invitation, but it's not quite finished. The party will last two hours. What time does it end?

Welcome to my party!

There will be pancake cake in the garden

10 o'clock – ________

It's going to be fun!

Findus after lunch

Findus has learned that there are 24 hours in a day. But clocks with hands only show 12 hours. Digital clocks show all 24 hours so after 12 noon (12.00) the next hour of 1 o'clock in the afternoon is shown as 13.00 and so on. Can you match the clocks to the correct times?

16:00

19:00

14:00

18:00

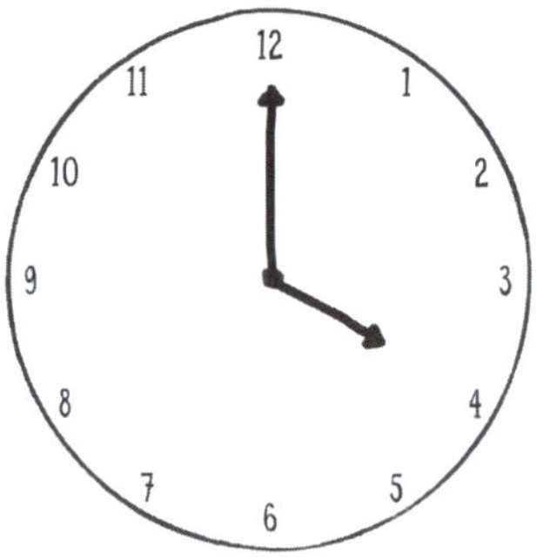

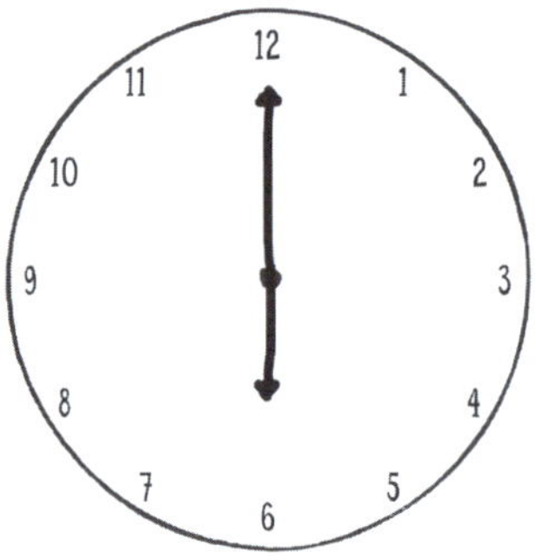

The hens' day

Pettson has made a daily schedule for the busy hens and pinned it on their wall. But he forgot to add the times! Can you suggest what times they may do the things on the list?

Eat breakfast

: ____ o'clock

Lay eggs

: ____ o'clock

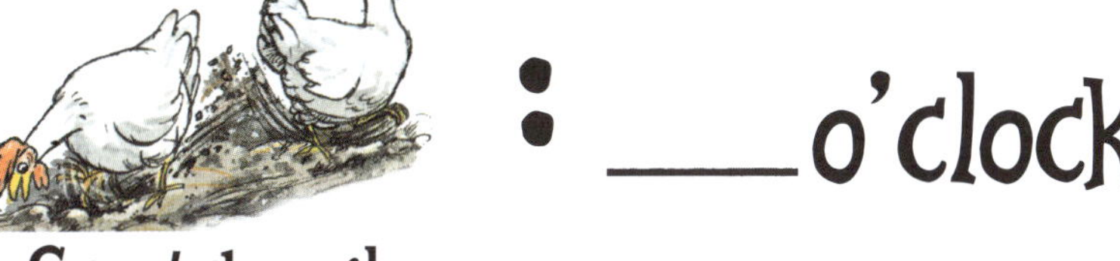

Scratch soil

: ____ o'clock

Go to bed

: ____ o'clock

Pettson's new clock

Gustavsson has given Pettson a new digital clock, but Pettson can't quite figure it out. Can you help him? Draw lines between the digital clocks and the clocks on the facing page that show the same times.

10.30

06.00

15.00

17.30

12
1
2
3
4
5
6
7
8
9
10
11

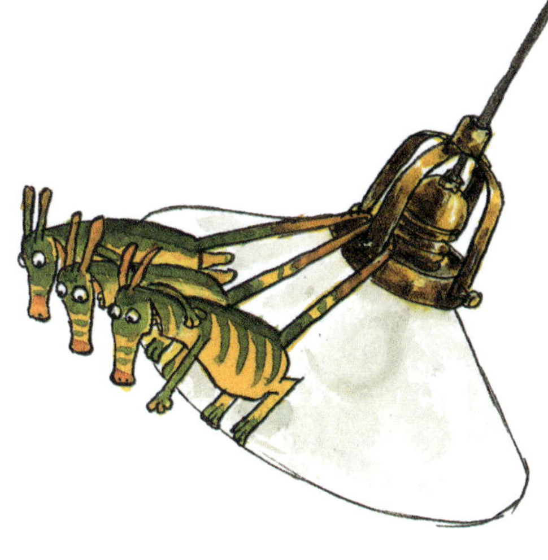

12
1
2
3
4
5
6
7
8
9
10
11

12
1
2
3
4
5
6
7
8
9
10
11

12
1
2
3
4
5
6
7
8
9
10
11

Telling the time: Analogue & digital
Night/Morning

1am = 01.00

2am = 02.00

3am = 03.00

4am = 04.00

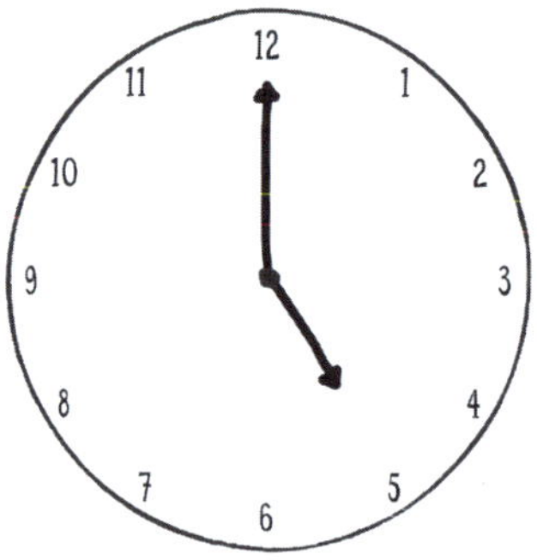

5am = 05.00

6am = 06.00

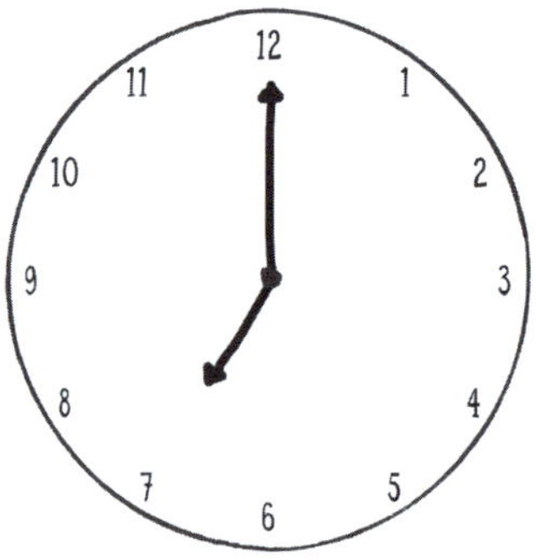

7am = 07.00

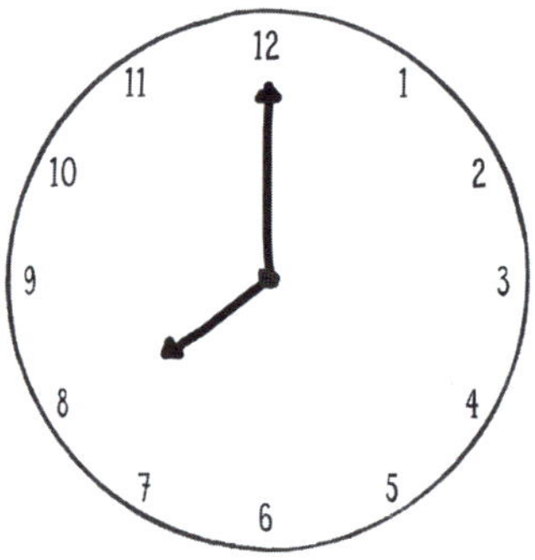

8am = 08.00

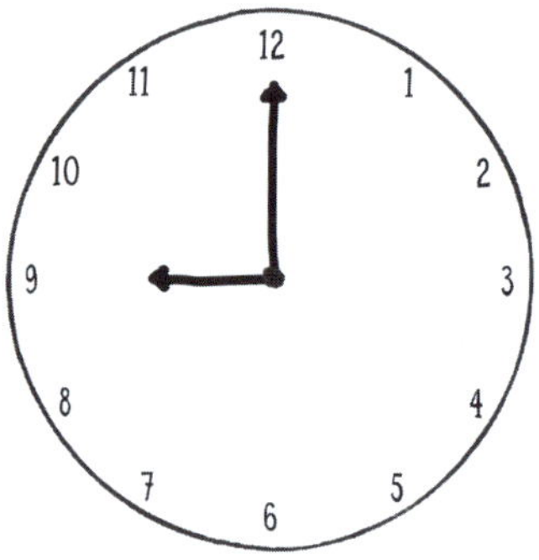

9am = 09.00

10am = 10.00

11am = 11.00

12am = 12.00

Telling the time: Analogue & digital
Afternoon/Evening

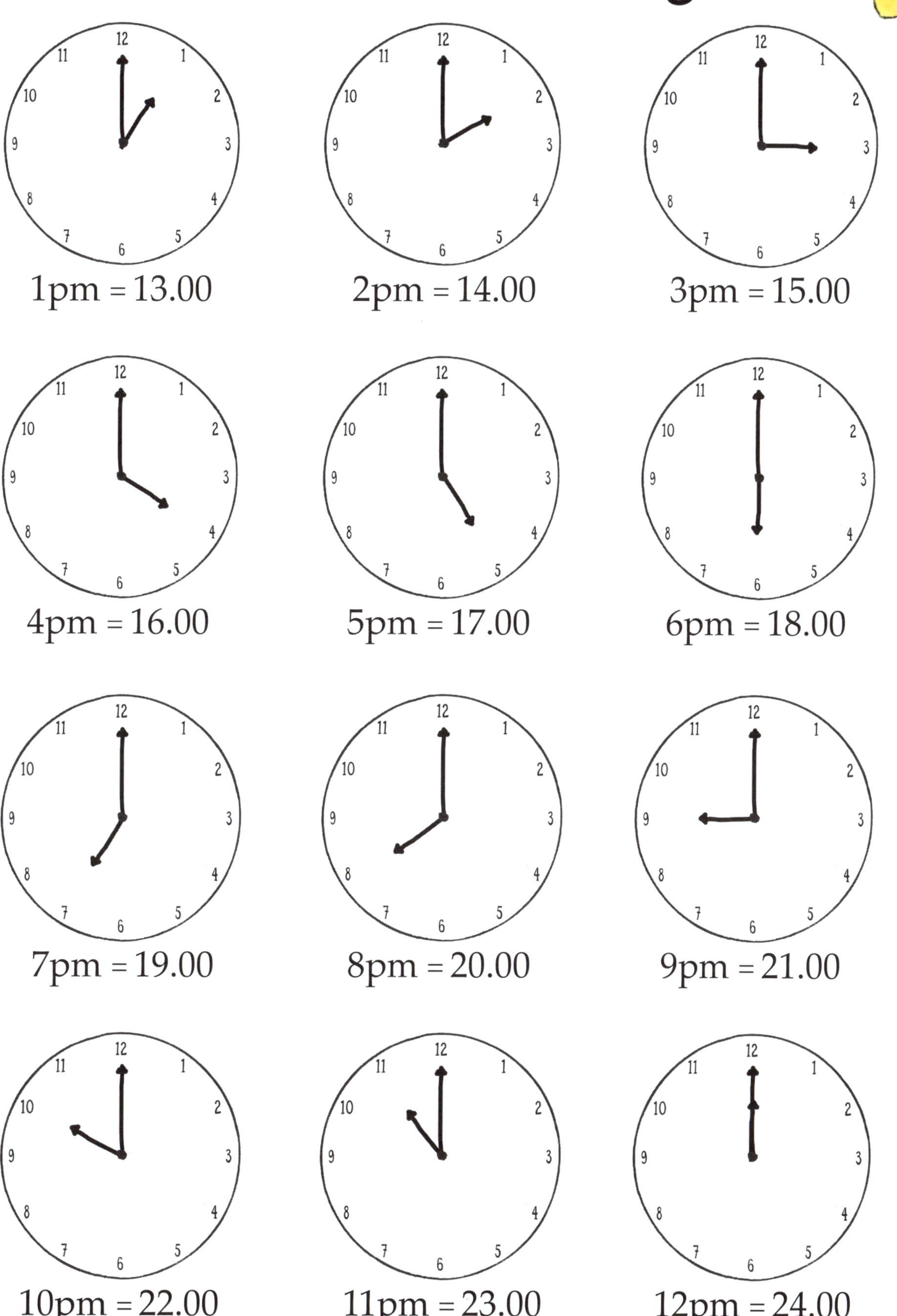

Answers

Pettson's schedule
Pettson likes to plan his day. Can you match the times on the clocks to the correct activities?
Have breakfast at 8 o'clock
Do the crossword at 11 o'clock
Tinker in the shed at 3 o'clock

Cock-a-doodle-doo!
Pavarotti the rooster's most important job is to wake everyone up at 7 o'clock. But he's not an early bird. What time does he start crowing on the following days?
Monday
8 o'clock
Tuesday
10 o'clock
Wednesday
12 o'clock

What time is it?
The muckles have been playing in the old clock and now it's stopped working. Can you help them by drawing on the missing hands to show the correct time?
9 o'clock
4 o'clock
2 o'clock
Half past ten
A quarter to five

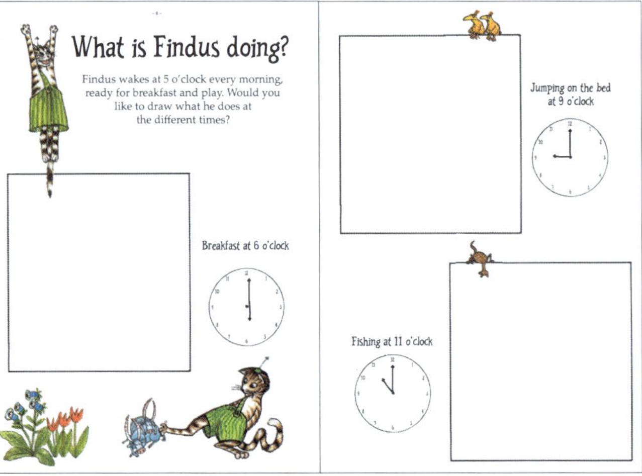
What is Findus doing?
Findus wakes at 5 o'clock every morning, ready for breakfast and play. Would you like to draw what he does at the different times?
Breakfast at 6 o'clock
Jumping on the bed at 9 o'clock
Fishing at 11 o'clock

Will Findus be late?
It is a quarter to two and Pettson will serve pancakes at 3 o'clock. Findus can choose three different ways to get home. Which path should he take to get back in time?
Two hours
One and a half hours
One and a quarter hours
X

Answers

The fox is coming!
The fox is on the prowl, so Findus and Pettson must make sure to lock up the hens at night. To do things properly they should take the following steps – but in what order?
Shut and bolt the door
3
4
Keep a lookout for the fox
1
Herd the hens into their coop
Put the hens to bed
2

Moving house
The muckles are sick of the rooster's crowing, so they decide to move to the other end of the garden. But their cart is too full and things keep falling off. Every hour two things fall off. How long have the muckles been travelling?
5 hours
What time is it?
The muckles have been at it again.
Draw the hands on the clocks to show the right times.
7 o'clock
Half past six

How long to go?
It is 5 o'clock in the morning. Findus wonders how many hours are left until the things in the pictures happen. What do you think?
2 hours
4 hours
12 hours

How long to go?
It is now 2 o'clock in the afternoon. Findus wants to know how long he must wait until the fun starts. Can you help him work it out?
2 hours
5 hours
8 hours

Muckle get-together
Two muckles plan to have a picnic under the big tree at 3 o'clock. The muckle in the tall house lives further away and it will take her one hour to reach the tree. It will only take her friend half an hour. What time should each of them leave home to arrive on time?

Answers

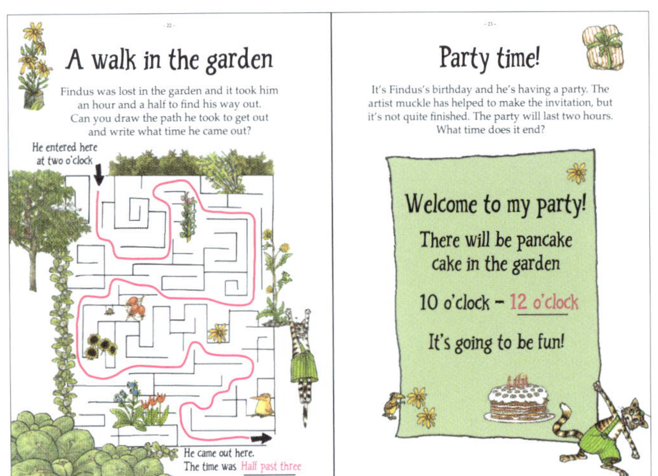

A walk in the garden

Findus was lost in the garden and it took him an hour and a half to find his way out. Can you draw the path he took to get out and write what time he came out?

He entered here at two o'clock

He came out here. The time was Half past three

Party time!

It's Findus's birthday and he's having a party. The artist muckle has helped to make the invitation, but it's not quite finished. The party will last two hours. What time does it end?

Welcome to my party!
There will be pancake cake in the garden
10 o'clock – 12 o'clock
It's going to be fun!

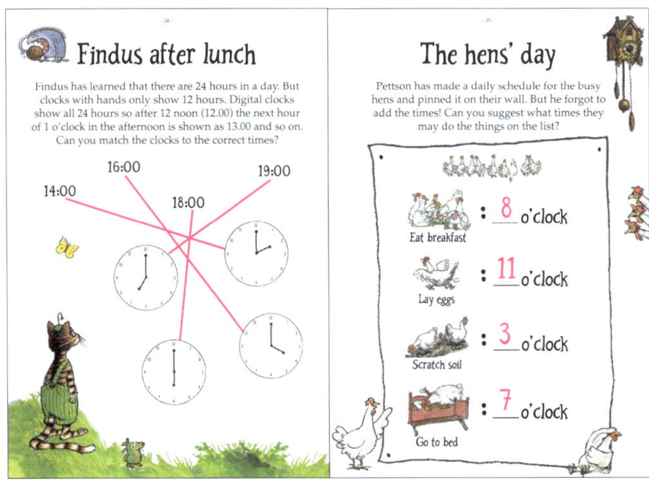

Findus after lunch

Findus has learned that there are 24 hours in a day. But clocks with hands only show 12 hours. Digital clocks show all 24 hours so after 12 noon (12.00) the next hour of 1 o'clock in the afternoon is shown as 13.00 and so on. Can you match the clocks to the correct times?

16:00
19:00
14:00
18:00

The hens' day

Pettson has made a daily schedule for the busy hens and pinned it on their wall. But he forgot to add the times! Can you suggest what times they may do the things on the list?

Eat breakfast : 8 o'clock
Lay eggs : 11 o'clock
Scratch soil : 3 o'clock
Go to bed : 7 o'clock

Pettson's new clock

Gustavsson has given Pettson a new digital clock, but Pettson can't quite figure it out. Can you help him? Draw lines between the digital clocks and the clocks on the facing page that show the same times.

10.30
06.00
15.00
17.30